Tamarind, the Indian Date

Growing Practices and Health Benefits

Roby Jose Ciju

All Rights Reserved. No parts of this publication may be reproduced, stored in a retrieval system, or transmitted, in any form or by any means, electronic, mechanical, photocopying, recording, or otherwise, without the prior permission of agrihortico

© 2014 AGRIHORTICO

CONTENTS

TAMARIND, THE INDIAN DATE 1

INTRODUCTION	1
ORIGIN AND DISTRIBUTION	2
TAXONOMY	2
BOTANICAL DESCRIPTION	3
PRODUCTION AREAS	4
SEASON	4
USES OF TAMARIND	5
INTRODUCTION	5
ORNAMENTAL USES	5
TAMARIND KERNEL POWDER	5
TIMBER USES	6
FUEL USES	6
CULINARY USES	6
OTHER USES	7
MEDICINAL USES AND HEALTH BENEFITS OF TAMARINDS	8
NUTRITION IN RAW TAMARINDS	9
NUTRITION IN CANNED TAMARIND NECTAR	11
MEDICINAL USES OF TAMARIND	12
TAMARIND GROWING PRACTICES	14
SITE SELECTION	14
CLIMATIC REQUIREMENTS	14
SOIL REQUIREMENTS	14
CULTIVARS AND VARIETIES	15
PROPAGATION	15
NURSERY PRODUCTION/RAISING OF SEEDLINGS	16

GRAFTING AND BUDDING	16
AIR LAYERING	16
PLANTING	16
STAKING OF YOUNG PLANTS	17
FROST INJURY	17
WATERING	17
FERTILIZER APPLICATION	17
INSECT PESTS	18
STORAGE PESTS	18
DISEASES	18
PRUNING	18
AFTERCARE	19
WEEDING	19
HARVESTING	19
YIELD	20
HOME STORAGE INSTRUCTIONS	20
STORAGE FOR PROCESSING PURPOSES	20
FOR LONG TERM STORAGE	21
INTERCROPPING IN A TAMARIND PLANTATION	21
ECONOMIC LIFE	21
FURTHER REFERENCES	**22**

TAMARIND, the Indian Date

Introduction

The word 'tamarind' has two words, 'tamar' and 'ind'. "Tamar" is an Arabic word means 'date' and "ind" is a hindi word means 'India'. Hence tamarind is popularly known as 'the Indian Date'. Tamarind is mainly grown for its edible fruit.

Scientific name of tamarind is *Tamarindus indica* (syns. T. occidentalis and T. officinalis). It belongs to the family Leguminosae (syns. Fabaceae or Caesalpinioideae). This leguminous tree is tropical in growth habit with a preference for dry semiarid climate. It is mainly grown for its pod-shaped edible fruits and tender leaves which are used as vegetables. Tamarind has some medicinal values as well. Tamarind wood may be used for timber and carpentry purposes. Some of the common names of tamarind are Tamr Hindi, Indian Date, Dakkar, Tamarindo, Tamarin, and Sampalok.

Tamarind should not be confused with "Malabar Tamarind" and "Manila Tamarind". Malabar tamarind is *Garcinina cambogia* while Manila tamarind is *Pithecellobium dulce*.

Origin and Distribution

It is native to tropical Africa but got naturalized in Indian subcontinent, West Indies and tropical Americas. Indian tamarind is more popular in the global market. American tamarind and West Indian tamarind also has a market.

Taxonomy

Kingdom	Plantae/Angiosperms
Class:	Equisetopsida
Subclass:	Magnoliidae
Superorder:	Rosanae
Order:	Fabales
Family:	Leguminosae/ Fabaceae/Caesalpinioideae
Genus:	Tamarindus
Species	Indica

Botanical Description

Tree	Tamarind is a slow-growing, evergreen, tropical, perennial tree with long drooping branches and dense open canopy. Under optimum conditions the tree reaches up to 80 feet high with a spread of 20 to 35 feet. In Africa and Asia, the tamarind tree reaches its maximum height of 80 feet while in Americas, the tree seldom reaches more than 15 to 25 feet in height.
Foliage/Leaves	Tamarind trees have bright, evergreen dense foliage. Foliage is comprised of alternately arranged pinnately compound feathery leaves and each leaf contains up to 10-20 leaflets. Each leaflet is about one inch long. The leaflet closes up during night.
Flowers	In tamarind tree, flowering is not spectacular event. However the tamarind does flower with reddish pinkish and yellowish elongated flowers. Actually flower colour is yellow. Flower buds appear red or pink because of the colour of the outer sepals which are red or pink in colour. When flower bud opens sepals fall off. Each flower is about one inch length.
Fruit	Fruit set may not happen in cool climate as tamarind trees need dry semi-arid climate for fruit development. Fruit is a pod which is dark brown in colour. The length of pod ranges from about 10 centimeters to 25 centimeters in length. Each pod is filled with soft brownish pulp and may contain 1 to 12 large flat seeds embedded in it. Fruit pulp is of great commercial value which is rich in sugar, tartaric acid, vitamin B and calcium.

Production Areas

Today, India is the largest producer of tamarind followed by USA. In USA, Florida is a major tamarind production area.

Season

Areas	Season
In USA and the West Indies	Flowering time: Summer Fruit set: December and January Fruit Ripening: April to June
In Hawaii Islands	Fruit Ripening: Late Summer and Fall
In India	Fruit Ripening and Harvesting: January to April

Uses of Tamarind

Introduction

Tamarind is a multipurpose tree. Timber of a fully grown tree yields softwood which is used for various carpentry purposes. Young, tender leaves are used as a delicious vegetable. Its pod-shaped fruit has great commercial value. Fruit may be eaten raw or may be used in various culinary or medicinal preparations. Fruit pulp extracted from mature, ripe fruits after removing seeds are available in the market for culinary uses.

Ornamental Uses

It is a perennial evergreen tree hence can be grown as an ornamental and garden tree. It has densely branched open canopy hence can be grown as a shade-giving avenue tree. In many tropical Asian countries, tamarind tree is used for making outdoor bonsai trees. In temperate countries, tamarind trees are best suitable for creating indoor bonsai.

Tamarind Kernel Powder

Seed kernels are ground to form a powder. It is available in the market as tamarind kernel powder (TKP) which is used as a cattle fodder along with other ingredients. However, some researchers suggest that TKP may have some health hazards such as exposure to finely ground powder may induce asthma and allergies to some people.

Tamarind kernel powder (TKP) is used as a sizing material in textile and leather industry. Jellose, a polysaccharide present in

TKP forms a gel with sugar concentrate which may be used as a substitute for fruit pectin.

Timber Uses

Tamarind heartwood is dark brown in color, very strong and durable. It is also insect-resistant and hence used for various carpentry uses such as furniture making and wood flooring.

Fuel Uses

Tamarind sapwood is pale yellow in color and is valued as firewood.

Culinary Uses

Tamarind pulp is used to make rasam, sambhar, tamarind juice, tamarind sauce, tamarind jam, tamarind rice, tamarind syrup, tamarind nectar, tamarind-flavored drinks, candies, chutneys and pickles. Young, tender leaves are used to prepare delicious food preparations with prawns, meat or pulses. Leaves are used as leafy vegetables also. Chopped leaves are used in soups.

Other Uses

Plant Part	Uses	Description
Fruit pulp	Metal Polish	Fruit pulp mixed with sea water may be used as a metal polish to clean silver, copper and brass utensils
Leaves	Fodder	Tamarind leaves are excellent fodder for cattle and goats. Tamarind leaves may also be used in sericulture as fodder for silkworms
Foliage	Biomulch	Foliage may be used as a biomulch in plantations, particularly tobacco plantations
Flowers	Source of Nectar	Flowers are filled with golden-yellow colored honey which is a good source of nectar for honeybees
Seeds	Textile Industry	TKP is used in textile industry for sizing and finishing cotton, jute and spun viscose, and for dressing homemade blankets
Seed Oil	Illuminant and Varnish	Seed oil is used as an illuminant and for varnishing dolls and idols. The oil is said to be palatable and of culinary quality
Twigs	Chew sticks	Tender twigs may be used as "chew sticks"
Bark	Tanning Hides and Dyeing	Bark contains up to 7% tannin and hence used in tanning hides and in dyeing. Barks may also be burned to make an ink

Medicinal Uses and Health Benefits of Tamarinds

Tamarind fruit pulp is rich in antioxidants and hence helps fight against cancer. Tamarind is rich in Vitamin C and hence protects against scurvy, a disease caused by Vitamin C deficiency. Pulp has several medicinal properties when given as infusion it treats bile disorders and febrile conditions. Tamarind fruit pulp which is sweetish and acidic in taste is used for serving curries, chutneys, sauces and soups.

Tamarind is an essential ingredient in many south Indian food preparations such as tamarind rice, sambhar, and rasam and tamarind chutney. Tender leaves, flowers and young seedlings of tamarind are consumed as a nutritious vegetable. Tamarind pulp is used for making nutrient-rich candies and jellies.

Nutrition in Raw Tamarinds

Nutrient	Unit	Value per 100 g	1 cup, pulp = 120.0g	1 fruit (3" x 1") = 2.0g
Proximates				
Water	g	31.4	37.68	0.63
Energy	kcal	239	287	5
Protein	g	2.8	3.36	0.06
Total lipid (fat)	g	0.6	0.72	0.01
Carbohydrate, by difference	g	62.5	75	1.25
Fiber, total dietary	g	5.1	6.1	0.1
Sugars, total	g	38.8	46.56	0.78
Minerals				
Calcium, Ca	mg	74	89	1
Iron, Fe	mg	2.8	3.36	0.06
Magnesium, Mg	mg	92	110	2
Phosphorus, P	mg	113	136	2
Potassium, K	mg	628	754	13
Sodium, Na	mg	28	34	1
Zinc, Zn	mg	0.1	0.12	0
Vitamins				
Vitamin C, total ascorbic acid	mg	3.5	4.2	0.1
Thiamin	mg	0.428	0.514	0.009
Riboflavin	mg	0.152	0.182	0.003
Niacin	mg	1.938	2.326	0.039

Vitamin B-6	mg	0.066	0.079	0.001
Folate, DFE	µg	14	17	0
Vitamin B-12	µg	0	0	0
Vitamin A, RAE	µg	2	2	0
Vitamin A, IU	IU	30	36	1
Vitamin E (alpha-tocopherol)	mg	0.1	0.12	0
Vitamin D (D2 + D3)	µg	0	0	0
Vitamin D	IU	0	0	0
Vitamin K (phylloquinone)	µg	2.8	3.4	0.1
Lipids				
Fatty acids, total saturated	g	0.272	0.326	0.005
Fatty acids, total monounsaturated	g	0.181	0.217	0.004
Fatty acids, total polyunsaturated	g	0.059	0.071	0.001
Cholesterol	mg	0	0	0
Other				
Caffeine	mg	0	0	0

Source: USDA Nutrient Database

Nutrition in Canned Tamarind Nectar

Nutrient	Unit	Value per 100 g	1 cup = 251.0g
Water	g	84.97	213.27
Energy	kcal	57	143
Protein	g	0.09	0.23
Total lipid (fat)	g	0.12	0.3
Carbohydrate, by difference	g	14.73	36.97
Fiber, total dietary	g	0.5	1.3
Sugars, total	g	12.7	31.88
Calcium, Ca	mg	10	25
Iron, Fe	mg	0.75	1.88
Magnesium, Mg	mg	4	10
Phosphorus, P	mg	2	5
Potassium, K	mg	27	68
Sodium, Na	mg	7	18
Zinc, Zn	mg	0.02	0.05
Vitamin C, total ascorbic acid	mg	7.1	17.8
Thiamin	mg	0.003	0.008
Riboflavin	mg	0.003	0.008
Niacin	mg	0.07	0.176
Vitamin B-6	mg	0.007	0.018
Folate, DFE	µg	1	3
Vitamin B-12	µg	0	0
Vitamin A, RAE	µg	0	0
Vitamin A, IU	IU	0	0
Vitamin E (alpha-tocopherol)	mg	0.12	0.3
Vitamin K (phylloquinone)	µg	0.1	0.3
Cholesterol	mg	0	0

Source: USDA Nutrient Database

Medicinal Uses of Tamarind

Medicinal use	Plant part used	Description
Refrigerants	Tamarind pulp	Tamarind pulp is used as refrigerants in fevers
Laxative	Tamarind pulp	It clears congestion in the bowels
Carminative	Tamarind pulp/Tamarind nectar	It soothes the digestive system
Digestive	Tamarind pulp/Tamarind nectar	It is a remedy for indigestion
Antiscorbutic	Tamarind pulp	It may be used as a substitute for lime juice
Anti-inflammatory	Tamarind pulp	It reduces swellings. In traditional medicines the pulp is applied on inflammations. In some parts of the world, a decoction prepared from tamarind pulp is used as a remedy in cases of gingivitis, asthma and eye inflammations
Liniment	Tamarind pulp	Tamarind pulp is used as liniment for rheumatism
Poultices	Bark, leaves and flowers, dried or boiled	Poultices made from the bark, leaves or flowers are applied on open sores, boils,

		swollen joints, sprains and caterpillar rashes
Antiseptics	Lotions and extracts	It destroys skin infections
Astringent	Bark of the tree/seed coat	It dries up sebum gland secretions
Health Tonic	Tamarind pulp	Tonics strengthen the whole body system
Tooth Health	Raw tamarind fruit, tamarind pulp and juice	Since tamarind is high in calcium content, it promotes teeth health

Tamarind Growing Practices

Here below is a brief description of growing practices for tamarinds.

Site Selection

Open field cultivation is recommended. Container growing is not suitable for tamarind as they subsequently grow into large trees. Open, spacious, sunny location is best suitable for growing tamarind plants.

Climatic Requirements

Semi-arid tropical climate is best for the commercial cultivation of tamarind. Tamarind can be grown in any areas where the temperature reaches 46°C maximum and 0°C minimum. Average rainfall requirement is 500–1,500mm. The optimum altitude required for tamarind cultivation is 1,000m above MSL (mean sea-level).

Soil Requirements

Tamarind can be grown in all types of soils provided that proper drainage is there. However, tamarind plants grow well in well-drained deep loamy or alluvial soils. Tamarind can be grown in poor soils such as rocky or porous soils too but soil fertility should be replenished time to time. Tamarind plants prefer slightly acidic soils and are tolerant to saline and alkali soils.

Cultivars and Varieties

Variety	Description
Indian Types	PKM 1:-Early variety; average yield is 263kg pods/tree; pulp content 39%.; can yield up to 26 tons of pods/ha if transplanted at a spacing of 10m × 10m Urigam:- A local variety; very long pods; sweet pulp Other Indian varieties:-Hasanur,Tumkur Prathisthan, DTS 1, Yogeshwari Indian types have longer pods with 6 - 12 seeds
Asian Types	Makham Waan is a popular variety in Thailand In Africa and Asia, the tamarind tree reaches its maximum height of 80 feet
American Types	Manila Sweet is a popular variety in USA in Americas, the tree seldom reaches more than 15 to 25 feet in height.
West Indies and African Types	the West Indian types have shorter pods containing only 3 - 6 seeds In Africa and Asia, the tamarind tree reaches its maximum height of 80 feet

Depending upon the presence of sugar content in the pulp, tamarinds may be categorised into sweet tamarinds and sour tamarinds.

Propagation

Seed propagation, layering, grafting and budding are practiced for propagating tamarind plants.

Nursery Production/Raising of Seedlings

Nursery beds are prepared and seeds are directly sown at a spacing of 20–25cm apart during March-April.

Seed germination may be enhanced by boiling them in hot water for a few seconds. Dipping seeds in 10 per cent cow urine or cow dung solution for one whole day just before sowing may protect them from insect-pest and fungal infestations in the field.

Irrigation is done soon after sowing and seeds start germinating within a week. Seedlings are regularly watered until they reach transplanting age. Seedlings may be raised in polythene bags also. Generally two year old seedlings are transplanted in the main field at the beginning of rainy season.

Grafting and Budding

Since true-to-type plants cannot be produced by seed propagation, grafting and budding are practiced for large scale propagation of tamarind plants. Approach grafting and Patch budding are proved to be quite successful.

Air Layering

Air layering may also be practised for propagating tamarind plants. In air layering, shoots treated with IBA (Indole Butyric Acid, a growth regulator)@4000 ppm.

Planting

Ideal time for planting tamarind seedlings in the main field is June–November. Pits of 1m × 1m × 1m size are dug in the field at a spacing of 10m × 10m. Seedlings are placed in the pit without disturbing the root ball and pit is then covered with a mixture of farmyard manure @ 15kg/pit and top soil.

Staking of Young Plants

Support is necessary for growing young plants until they are strong enough to support themselves.

Frost Injury

Growing tamarind trees are very susceptible to frost and hence need to be protected from severe frost. Large mature trees are very hardy.

Watering

Regular watering @once in 7 days is done until plants get established in the field. Established tamarind plantation may not require regular watering since tamarind is a plant of semi-arid regions. Hence tamarind trees can withstand drought conditions fairly well. Overwatering tamarind plantation must be avoided at any cost as the growing plants may not survive well in soggy waterlogged soils.

Fertilizer Application

Established tamarind plantations may not require regular fertilizer application. However young growing trees should be fertilized regularly until they get established in the field. According to some researchers, the recommended rate of nitrogen, phosphorous and potassium requirement in countries such as India is @200:150:250 g of NPK per tree per year. Application of 25 kg of FYM (farm yard manure) and 2 kg of Neem cake per tree per year along with NPK fertilizers enhances productivity and production.

Insect Pests

Leaf caterpillar (Achaea janata) and Storage beetle (Pachymeres gonagra) may be a problem. But these can effectively be controlled by the application of a suitable insecticide. Other insects found affecting tamarind plantations in India are mealybugs, aphids, white flies, thrips and a variety of scales.

Storage Pests

Different kinds of weevils and fruit borers are found infesting the ripening pods on the trees as well as stored fruits.

Diseases

Tamarind plants may be susceptible to a fungal infection called powdery mildew which can be controlled by spraying a suitable fungicide at the recommended rate.

Pruning

Two types of pruning are practised in tamarind plantation. There is pruning of young growing tamarind trees and then pruning of established plants which is called maintenance pruning. While pruning young plants, 3-5 well-formed healthy branches are allowed on the young, growing tamarind tree so that a well-structured canopy is developed over a period of time. Maintenance pruning is required to remove all distorted, diseased and unhealthy branches from the established trees. Pruning a growing tamarind plant is essential for maintaining optimum production throughout its economic life.

Aftercare

Established tamarind plantation may not require regular care. Growing tamarind plants need regular care. Pruning is an ongoing procedure. All unwanted growth and sprouts are removed as and when required.

Weeding

There are hardly any weeds that are found affecting tamarind plants. Weeding may be necessary during initial growth of the plants. Established tamarind plantation hardly requires any weeding.

Harvesting

Seedling plants start yielding in 6-8 years after planting. Grafts and budded plants start yielding from 4th year onwards. Harvesting in India and surrounding regions may be done during January–April. In USA and West Indies fruits ripen during April to June and harvesting may be done during this period.

Ripened pods stay on the tree even after 6 months of reaching ripening stage. In India harvesting may be done by shaking the branches to make the ripened pods fall off naturally from the trees or the grower waits until the ripened pods fall off naturally from the trees and then gather them from the ground. Harvesting with poles is not recommended as it will damage flowers, immature fruits and tender leaves.

For processing purposes, harvesting may be done by gently pulling the pods away from the stalk attached to it.

For fresh markets, pods may be harvested along with stalks by clipping them from the branches.

Yield

Average yield is 26 tons of pods/ha per year in commercial production practices. It is estimated that a well-grown healthy tamarind tree annually produces approximately 330 to 500 lbs (150-225 kg) of fruits.

Fruit Constituents

Fruit Pulp	30 to 55%
Shells and Fiber	11 to 30 %
Seeds	33 to 40%

Home Storage Instructions

Pods are shelled, pulp is extracted after removing seeds and then pulp is layered with sugar. This pulp-sugar mixture is then pressed into tight balls and covered with a white muslin cloth and kept air tight containers in a cool, dry place. Sugar is used as a preservative here.

Storage for Processing Purposes

Pods are shelled, pulp is extracted after removing seeds and then pulp is layered with sugar in barrels and covered with boiling syrup. Sugar is used as a preservative here.

For Long Term Storage

For long term storage of tamarind pulp, the layers of pulp are first steamed and then sun-dried for several days.

Intercropping in a Tamarind Plantation

Since a tamarind plantation comes in yield only after four years of planting, intercropping may be practised during initial four years. Peas and beans, seasonal vegetables, leafy vegetables, sesamum and sorghum may be considered as suitable intercrops.

Economic Life

A healthy well-grown tamarind plant has an economic life of up to 50 years. After that the tree continues to live, in some cases up to 150 years, but production and productivity declines considerably.

Further References

Burkill, H.M. (1995). *The Useful Plants of West Tropical Africa*. Royal Botanic Gardens, Kew.

DuPuy, D.J., Labat, J. -N., Rabevohitra, R, Villiers, J. -F., Bosser, J. & Moat, J. (2002). *The Leguminosae of Madagascar*. Royal Botanic Gardens, Kew.

Lewis, G., Schrire, B. Mackinder, B. & Lock, J. M. (eds) (2005). *Legumes of the World*. Royal Botanic Gardens, Kew.

Morton, Julia F. (1987). *Fruits of Warm Climates*. Creative Resources Systems, Inc. 1987. pp. 115-121.

Popenoe, Wilson. (1974). *Manual of Tropical and Subtropical Fruits*. Hafner Press. Facsimile of the 1920 edition. pp. 432-436.

www.ingramcontent.com/pod-product-compliance
Lightning Source LLC
Chambersburg PA
CBHW071756200526
45167CB00018B/2206